GERMANY'S TIGER TANKS
TIGERS AT THE FRONT

Also by the Author
GERMANY'S TIGER TANKS: Vol.1 - D.W. to TIGER I
GERMANY'S TIGER TANKS: Vol.2 - V.K.45.02 to TIGER II
GERMANY'S TIGER TANKS: Vol.3 - TIGER I & II: COMBAT TACTICS
GERMANY'S PANTHER TANK: THE QUEST FOR COMBAT SUPREMACY
TANK COMBAT IN NORTH AFRICA
PANZERTRUPPEN: 1933-1942
PANZERTRUPPEN: 1943-1945

GERMANY'S TIGER TANKS
TIGERS AT THE FRONT

A PHOTO STUDY COMPILED BY
Thomas L. Jentz

Schiffer Military History
Atglen, PA

Book design by Robert Biondi.

Copyright © 2001 by Thomas L. Jentz.
Library of Congress Catalog Number: 00-110568.

All rights reserved. No part of this work may be reproduced or used in any forms or by any means – graphic, electronic or mechanical, including photocopying or information storage and retrieval systems – without written permission from the copyright holder.
"Schiffer," "Schiffer Publishing Ltd. & Design," and the "Design of pen and ink well" are registered trademarks of Schiffer Publishing, Ltd.

Printed in China.
ISBN: 0-7643-1339-8

We are always looking for people to write books on new and related subjects. If you have an idea for a book, please contact us at the address below.

Published by Schiffer Publishing Ltd. 4880 Lower Valley Road Atglen, PA 19310 Phone: (610) 593-1777 FAX: (610) 593-2002 E-mail: Schifferbk@aol.com. Visit our web site at: www.schifferbooks.com Please write for a free catalog. This book may be purchased from the publisher. Please include $3.95 postage. Try your bookstore first.	In Europe, Schiffer books are distributed by: Bushwood Books 6 Marksbury Ave. Kew Gardens Surrey TW9 4JF England Phone: 44 (0)208 392-8585 FAX: 44 (0)208 392-9876 E-mail: Bushwd@aol.com. Free postage in the UK. Europe: air mail at cost. Try your bookstore first.

GERMANY'S TIGER TANKS
TIGERS AT THE FRONT

Tiger! – the very name that Allied troops feared. It came to symbolize the superiority of German tank design. No tank that the Allies fielded in World War II was comparable to this combination of the 88 mm gun with massive armor protection. Allied tankers didn't think that they stood a chance of defeating these formidable Tigers. They only hoped that these heavy tanks would breakdown, become immobilized in soft ground, or be damaged by lucky hits on vulnerable points.

Germany started development of heavy Panzer designs in 1937. They were named Durchbruchswagen (D.W. – breakthrough tank) and VK 30.01 (first model of a fully tracked tank in the 30 ton class). Only a few prototypes of these early heavy tanks were produced. An effort to accelerate the heavy Panzer program was first emphasized in a meeting with Hitler on 26 May 1941. This was before the invasion of Russia on 22 June 1941. Therefore, design of the Tiger I was not initiated as a response to the Russian T34 or KW tanks. Instead, the main concerns addressed during the meeting on 26 May 1941 were the problems of successfully combating British tanks and anti-tank guns. However, after encounters with the T34 and KW tanks, the design and production of an effective heavy Panzer were pursued with increased urgency.

Emphasis was now placed on thick armor protection and a very powerful gun capable of penetrating enemy heavy tanks at long range. These heavy Tiger tanks so impressed the Allied tankers in World War II that modern tanks are still based on these same criteria.

TIGER I DEVELOPMENT AND DESCRIPTION

The Tiger I was quickly designed utilizing components that had been invented and partially tested in previous heavy Panzers. The components for the Tiger I chassis had been mainly invented for the 30 and 36 ton class of heavy Panzer in the D.W. series from Henschel. The gun and turret for the Tiger I were designed by Krupp for mounting on a Porsche chassis.

During the meeting on 26 May 1941, a major design breakthrough was achieved as a result of the decision to install submerged fording equipment. This decision released the designers from the weight restrictions imposed by bridges. Freed from bridge weight restrictions, the designers could specify heavier armament and armor protection, which drastically increased the weight of the vehicle to 45 metric tons. However, the designers still had to meet the restrictive specifications governing the maximum dimensions for rail transportation.

The powerful 8.8 cm Kw.K.36 L/56 gun was mounted in the turret designed by Krupp. Secondary armament was provided by an M.G.34 mounted coaxially to the right of the main gun. Later, a second M.G.34 could be mounted on the cupola ring for anti-aircraft defense.

The turret had a unique horseshoe shape, with the open end covered by the 100 mm thick gun mantle. Armor thickness was 80 mm for the turret walls and 25 mm for the turret roof. Access to the turret was provided through a hatch in the cupola and a hatch directly over the loader's position. An exhaust fan was mounted behind the loader's hatch on the turret roof.

Adequate vision devices provided all-round viewing by the crew in the turret. The gunner had a binocular Turmzielfernrohr 9b sighting telescope with 2.5x magnification and a vision block to his left. The loader had a vision block to the right front and a pistol port to the right rear. The commander had all-round vision blocks in the cupola and a pistol port to the left rear.

The new chassis for the VK 45.01 (H) was created by altering the hull design for the VK 36.01. The superstructure sides were extended out over the tracks to create panniers, limited in their width due to restrictions for rail transport. The width of these side extensions was based on the area needed to house the radiators. The radiators were relocated to positions on both sides so that the center engine compartment could be sealed leak-tight for deep fording. There was a large, hinged rectangular hatch over the engine compartment. Unlike previous designs where the superstructure was bolted to the hull along a flange, the superstructure was welded to the hull. Access for maintenance of the engine, cooling system, and fuel system was accomplished by unbolting the sections of the rear deck.

The driver's front plate was 100 mm at 9 degrees, front nose plate 100 mm at 25 degrees, superstructure side plates 80 mm at 0 degrees, hull side plates 60 mm at 0 degrees vertical, tail plate 80 mm at 9 degrees, deck plates 25 mm at 90 degrees horizontal, and belly plate 25 mm horizontal.

A total of 92 rounds of ammunition were carried for the main gun. Sixty-four rounds were stored horizontally in covered bins in the panniers along the superstructure sides. Sixteen rounds were located in closed bins along the hull sides, six in a closed bin under the turret floor, and six in a closed bin beside the driver.

Direct vision was provided for the driver by a visor mounted in the front plate. For additional protection, the visor could be closed and the driver could use the Fahrerfernrohre (twin driver's periscopes). A fixed periscope in the driver's hatch provided the driver with a view toward the left front. The radio operator had a Kugelzielfernrohr 2 (sighting telescope) to aim the ball-mounted M.G.34 and a fixed periscope in the hatch above his head.

The drive train consisted of a high performance Maybach HL 210 P45, 12 cylinder motor delivering 650 metric HP at 3000 rpm, through an 8 speed Maybach Olvar 40 12 16 transmission onto the Henschel L 600 C double radius steering gear and final drives, designed to provide a maximum speed of 45.4 kilometers per hour. Maintaining a transverse torsion bar suspension, the combat weight of 57 metric tons was distributed over eight sets of interleaved 800 mm diameter rubber-tired roadwheels per side. The unlubricated 725 mm wide cross-country tracks provided an acceptable ground pressure (when the tracks sank to 20 cm) of 0.735 kilogram per centimeter squared.

The first Tiger I from the test series of three was completed by 20 April 1942 for a demonstration on Hitler's birthday. The first Tiger from the production series was completed and sent to Kummersdorf in May 1942 for testing. Altogether, a total of 1349 Tiger I were completed by Henschel – 3 in the test series and 1346 in the production series from April 1942 to August 1944.

TIGER II DEVELOPMENT AND DESCRIPTION

Henschel initiated an expedited program in November 1942 to develop an upgraded chassis for a turret housing the more powerful 8,8 cm Kw.K.43 L/71 gun. Originally known as the Tiger H3, Henschel was urged to complete the design and get production started as quickly as possible, as a replacement for the Tiger I with the shorter 8,8 cm Kw.K. L/56 gun.

The new hull design for the VK 45.03 consisted of sloping plates for increased protection. The front glacis plate was 150 mm at 50 degrees, front nose plate 100 mm at 50 degrees, superstructure side plates 80 mm at 25 degrees, hull side plates 80 mm at 0 degrees vertical, tail plate 80 mm at 30 degrees, deck plates 40 mm at 90 degrees horizontal, and front belly plate 40 mm horizontal and rear belly plate 25 mm horizontal. A large rectangular cover plate, flush with the roof in front of the turret, could be lifted to remove the transmission and steering units for maintenance without having to remove the turret. There was a large, hinged rectangular hatch over the engine, and the entire rear deck could be removed for maintenance of the engine, cooling system and fuel system.

48 rounds of ammunition for the main gun were stored horizontally in panniers on each side of the hull. The rounds were stowed in three groups (6, 7 and 11 rounds) on both sides. Each group was separated by 20 mm plates and covered by sliding metal panels. An additional 10 to 16 rounds were stacked loose on the turret floor.

A rotating periscope was provided for the driver to use when buttoned up. His seat, steering controls and accelerator pedal were adjustable to allow freedom and ease of driving with the head protruding from the open hatch. The radio operator had a Kugelzielfernrohr 2 (sighting telescope) to aim the ball-mounted M.G.34 and a periscope in the roof fixed at an angle of 16 degrees to the right front.

The drive train consisted of a high performance Maybach HL 230 P30, 12 cylinder motor delivering 750 metric HP at 3000 rpm, through an 8 speed Maybach Olvar 40 12 16 B transmission onto the Henschel L 801 double radius steering gear and final drives, designed to provide a maximum speed of 41.5 kilometers per hour. Maintaining a transverse torsion bar suspension, the combat weight of 68.5 metric tons was distributed over nine sets of overlapping 800 mm diameter steel-tired, rubber-cushioned roadwheels per side. The unlubricated 800 mm wide, double link combat tracks provided an acceptable ground pressure (when the tracks sank to 20 cm) of 0.76 kilogram per centimeter squared.

The turret mounted on the first Tiger II had been designed and produced by Krupp for mounting on a chassis designed by Porsche. However, due to technical problems, series production of the Porsche chassis was canceled, resulting in 50 turrets being available for mounting on the Tiger II. This turret was purposely designed with the front extended forward in order to move the gun trunnions forward. Thereby adequate room was gained between the gun breech and the turret race to allow for gun recoil, spent cartridge ejection and loading the long main gun rounds. Along with the sloping sides and roof, the rounded, 100 mm thick turret front presented a very poor target for enemy gunners. To aid in balancing the turret, which was necessary for reduction in the power needed to traverse the turret when the tank wasn't level, the rear of the turret was extended to act as a counterweight. Secondarily, this extra space in the turret rear provided easy access to main gun rounds stored in ready racks. The loader's task of maneuvering the long rounds within the cramped confines of the turret was simplified by pointing the nose of the ammunition toward the gun breech.

To increase the effective armor protection, the 80 mm thick turret sides were sloped inward at an angle of 30 degrees from the vertical. This decreased the width of the turret roof, necessitating the incorporation of a bulge in the left side to accommodate the commander's periscope cupola. The commander had a pivoting hatch in the cupola, the loader had a hatch directly overhead, and an escape hatch countersprung by torsion bars was provided in the turret rear. The rear of the turret could be unbolted to allow removal of all the internal components including the gun without dismounting the turret itself.

Adequate vision devices were provided with a binocular Turmzielfernrohr 9b/1 sighting telescope for the gunner, a fixed periscope for the loader, and all-round periscopes in the cupola for the commander. The pistol port with plug and a spent cartridge case ejection port, originally cut into the left turret side, were welded shut and covered with a Zimmerit anti-magnetic coating before leaving the assembly factory. The spent cartridge case ejection port was relocated to the turret roof behind the armored guard for the exhaust fan. The need for the pistol port was eliminated with the addition of the Nahverteidigungswaffe (close defense weapon) mounted on the turret roof. With 360 degrees traverse, the Nahverteidigungswaffe could be used to fire grenades, smoke and signal rounds. Secondary armament was provided by an M.G.34 mounted coaxially to the right of the main gun. A second M.G.34 could be mounted on the cupola ring for anti-aircraft defense.

A new turret designed by Krupp was introduced starting with the 51st Tiger II. Known as the "Serienturm" (series turret), it had a 180 mm thick front plate, 80 mm sidewalls, and 40 mm roof. The gun mantle was specifically designed to be immune to attack or being jammed. This design did not create high explosive blast pockets and prevented deflection of projectiles down onto the deck. By removal of the new hatch in the turret rear, all of the components inside the turret, including the gun and cradle, could be disassembled and removed without dismounting the turret itself. The shallower angle of the sidewalls allowed more room for storage of ammunition in the ready racks in the turret rear, 22 rounds versus 16 in the older design. Components on the turret roof remained the same as before.

The three Tiger II in a test series were completed by Henschel in November and December 1943. The first three Tiger II from the production series were completed in January 1944. Altogether, a total of 492 Tiger II were completed by Henschel – 3 in the test series and 489 in the production series from November 1943 to March 1945.

TIGER UNIT ORGANIZATION

In 1942, nine Tigers were issued along with 10 Pz.Kpfw.III to each schwere Kompanie (heavy company) along with 10 Pz.Kpfw.III. There were two schwere Kompanien in each independent schwere Panzer-Abteilung (heavy tank battalion). Two command Tigers and five Pz.Kpfw.III were issued to the headquarters company of each schwere Panzer-Abteilung. A total of five schwere Panzer-Abteilung were formed and sent to the front as follows:

- schwere Panzer-Abteilung 501 sent to Tunisia
- schwere Panzer-Abteilung 502 with one company sent to northern Russia, south of Leningrad, and one company to southern Russia
- schwere Panzer-Abteilung 503 sent to southern Russia
- schwere Panzer-Abteilung 504 sent to Tunisia and Sicily
- schwere Panzer-Abteilung 505 sent to Heeres Gruppe Mitte (army group middle) in central Russia

In addition, a schwere Kompanie with nine Tigers and 10 Pz.Kpfw.III was assigned to the Panzer-Regiments in Infanterie-Division "Grossdeutschland" (an army unit) and three SS Divisions (LSSAH, Das Reich, and Totenkopf) and sent to southern Russia early in 1943.

Starting in the Spring of 1943, the unit organization was changed to 14 Tigers in each schwere Kompanie and three command Tigers in the headquarters company for a total of 45 in each schwere Panzer-Abteilung. Those units already on the Eastern Front were sent Tigers to fill them to the new organization strength of 14 Tigers per company. The 501st and 504th, wiped out in Tunisia and Sicily, were later reformed with the 501st sent to Russia in the Fall of 1943 and the 504th sent to Italy in June 1944.

Five additional independent schwere Panzer-Abteilungen were formed for the army as follows:

- schwere Panzer-Abteilung 506 sent to southern Russia
- schwere Panzer-Abteilung 507 sent to Russia
- schwere Panzer-Abteilung 508 sent to Italy
- schwere Panzer-Abteilung 509 sent to southern Russia
- schwere Panzer-Abteilung 510 sent to Russia

In addition, the Tiger unit with "Grossdeutschland" was expanded to a full schwere Panzer-Abteilung with three schwere Kompanien in the Spring of 1943. Three schwere Panzer-Abteilungen were also formed for the SS in 1943 and initially numbered 101, 102, and 103. Two of these units, the 101st and 102nd, were sent to France in June and July 1944 where they were wiped out in the Falaise Pocket in August 1944. The last unit completely outfitted with Tiger I was schwere (FKl) Panzer-Abteilung 301 (heavy radio-controlled demolition vehicle battalion), which was sent to the Western Front in the Fall of 1944.

Other Tiger units sent to France were schwere Panzer-Abteilung 503 and a schwere (Fkl) Kompanie attached to the Panzer-Lehr Division. These two units were the first to employ the Tiger II in combat. Additional Tiger II were sent to the Western Front with the rebuilt 3.Kompanie/schwere Panzer-Abteilung 503 and 3.Kompanie/schwere SS-Panzer-Abteilung 101.

Tiger IIs were then issued to rebuilt schwere Panzer-Abteilungen and sent to the front as follows:

- schwere Panzer-Abteilung 501 to the Eastern Front
- schwere Panzer-Abteilung 504 to the Eastern Front
- schwere Panzer-Abteilung 506 to the Western Front
- schwere Panzer-Abteilung 503 to the Eastern Front
- schwere SS-Panzer-Abteilung 501 to the Western Front
- schwere Panzer-Abteilung 509 to the Eastern Front
- schwere SS-Panzer-Abteilung 502 to the Eastern Front
- schwere SS-Panzer-Abteilung 503 to the Eastern Front
- schwere Panzer-Abteilung 507 for defense of the homeland

There were a few other minor units with a few Tigers which were created by activating training units as the Allied advances brought the front to them. This last ditch effort was completely futile as an attempt to stop the overwelming numbers of Allied tanks at the end of the war.

ADDITIONAL FACTS AND DATA

This book has been created as a photo study showing the best available and rare photographs of Tigers with units at the front. Additional detailed information on the Tigers, based solely on research from original documents, is available from the publisher in the following books by this same author:

Germany's Tiger Tanks – D.W. to Tiger I – contains everything available on the development of heavy tanks by Henschel up to the Tiger I, including detailed descriptions, illustrations, photographs, and drawings showing all of the modifications in sequence as they occurred in the Tiger I production series.

Germany's Tiger Tanks – V.K.45.02 to Tiger II – describes in detail the development of the Tiger II by Henschel and has complete descriptions, illustrations, photographs, and drawings showing all of the modifications in sequence as they occurred in the Tiger II production series.

Germany's Tiger Tanks – Combat Tactics – contains everything one needs to understand the tactical employment of the Tigers in combat. It has all of the technical detail on the operational characteristics, armor protection, weapons accuracy, and armor-penetration ability of the main guns. Tactical manuals used by the Tiger units have been translated in full. Numerous original combat accounts written directly after the actions provide the only accurate stories of what it was like to fight with Tigers.

TIGERS IN TUNISIA AND SICILY

Schwere Panzer-Abteilung 501 was sent to Tunisia starting in November 1942 with 20 Tigers with their long 8.8 cm Kw.K.36 L/56 guns and 25 Pz.Kpfw.III Ausf.N with the short 7.5 cm Kw.K. L/24 guns. This is one of the two Tigers in the 4.Zug (4th platoon) of the 1.Kompanie. (BA)

Improvised heat guards surround the exhaust mufflers on one of the two Tigers in the 1.Zug (platoon) of the 1.Kompanie of schwere Panzer-Abteilung 501. (BA)

OPPOSITE: During December 1942 and January 1943, the schwere Panzer-Abteilung 501 fought for control of the mountain passes in northern Tunisia. Here tactical number 121 (platoon leader of the 2.Zug of the 1.Kompanie) is guided across a ford. (BA)

The first Tiger (Fgst.Nr.250011, tactical number 231) captured by the British. Both the British and Germans took credit for blowing it up with explosive charges the next night. (NA)

In February 1942, schwere Panzer-Abteilung 501 was incorporated into Panzer-Regiment 7 of the 10.Panzer-Division. As shown by the tactical number 732, the 1.Kompanie/schwere Panzer-Abteilung 501 was renumbered as the 7.Kompanie/Panzer-Regiment 7. (BA)

THIS PAGE AND OPPOSITE: Two of the seven Tigers (tactical numbers 823 and 833 of the 8.Kompanie/Panzer-Regiment 7) lost during Operation "Ochsenkopf". They were loaded with explosive charges and blown up to prevent them from falling intact into enemy hands. (NA)

Only part of schwere Panzer-Abteilung 504 with 11 Tigers and 19 Pz.Kpfw.III Ausf.M with long 5 cm Kw.K.39 L/60 guns was sent to Tunisia in March and April 1943. This Tiger (Fgst.Nr.250122, tactical number 131, produced in February 1943) was captured by the British and returned to England. It is now being restored to running condition by The Tank Museum at Bovington in Dorset. (TTM)

The other half of schwere Panzer-Abteilung 504 was sent to Sicily. This is one of ten Tigers from the 2.Kompanie that were lost in the first three days in action following the Allied landing on 10 July 1943. (NA)

Another Tiger from the 2.Kompanie/schwere Panzer-Abteilung 504 on Sicily which was abandoned after attempting to reverse after breaking the right track (possibly on a mine). (TTM)

TIGER I ON THE EASTERN FRONT

The 1.Kompanie/schwere Panzer-Abteilung 502 was the first unit to use Tigers in combat on 16 September 1942. This is the first Tiger captured by the Russians in January 1943. Tactical number 100 (designating the company commander's Tiger) is stenciled on the improvised stowage bin and the "Mammut" (the unit emblem for the 502nd) is stenciled on the turret rear. (SF)

Schwere Panzer-Abteilung 503 was sent to southern Russia in December 1942. This Tiger (tactical number 123) in the 1.Kompanie has been partially covered with whitewash as winter camouflage. (BA)

A hit by a Russian anti-tank gun (probably 76.2 mm) which failed to penetrate the 100 mm thick driver's front plate of this Tiger. (BA)

OPPOSITE: The turret had to be removed in order to replace the transmission or steering gear in the Tiger. Many of the transmissions in Tigers belonging to schwere Panzer-Abteilung 503 failed due to design deficiencies and inexperienced drivers. (BA)

Maintenance personnel assigned to the Werkstatt-Zug (workshop platoon) working on a disassembled transmission from a Tiger. (SB)

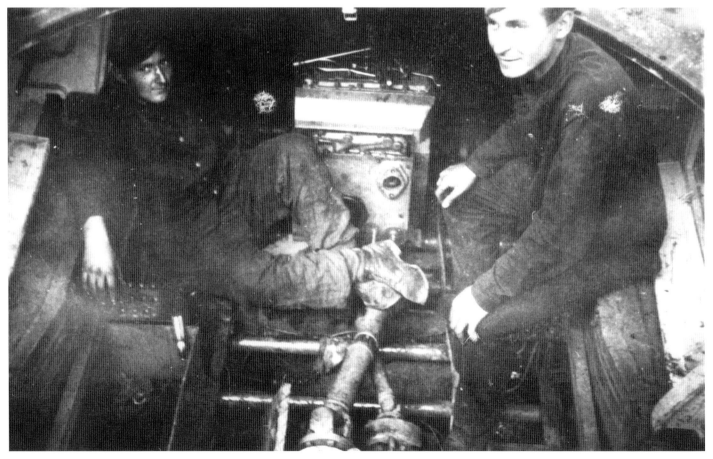
The turret has been removed and the main drive shaft disconnected from the transmission in this Tiger. (SB)

The first 20 Tigers in schwere Panzer-Abteilung 503 still had Pz.Kpfw.III style stowage bins on the turret. Tiger (tactical number 201), with roadwheels removed to replace rubber tires, had two pistol ports on the turret rear. (SB)

A hit from a Russian anti-tank gun (probably 76.2 mm) which failed to penetrate the 100 mm thick frontal hull armor. (SB)

Tigers from the 3.Kompanie/schwere Panzer-Abteilung 503 can be identified by the two national crosses stenciled on the large improvised stowage bin on the turret rear. (WR)

The large portal crane belonging to the Werkstatt-Zug of schwere Panzer-Abteilung 503 has been camouflaged with tree branches. (SB)

A 6 ton crane mounted on an 18 ton Zugkraftwagen (Sd.Kfz.9/1) was used to lift off the rear deck and remove the Maybach HL 210 P45 engine from this Tiger of schwere Panzer-Abteilung 503. (SB)

A replacement Tiger produced in March/April 1943 which arrived at the front for schwere Panzer-Abteilung 503 by the end of April 1943. (SB)

A Tiger (tactical number 114 of schwere Panzer-Abteilung 503) being towed by an 18 ton Zugkraftwagen (Sd.Kfz.9). (BA)

This Tiger (tactical number 114) was one of 20 replacements sent to schwere Panzer-Abteilung 503 in April 1943. (BA)

PAGES 34-39: Multiple views of Tiger (tactical number 123 of schwere Panzer-Abteilung 503) in the Spring of 1943. Produced in December 1943, it still had two pistol ports on the turret rear and a Pz.Kpfw.III stowage bin. Both capped (darker color) armor-piercing shells and fuzed (lighter color) high-explosive shells are being loaded. (BA)

The spare track links held by a bar across the front hull of this Tiger (tactical number 323 belonging to the 3.Kompanie/schwere Panzer-Abteilung 503) was a field modification accomplished by the Werkstatt-Zug (maintenance platoon). (BA)

OPPOSITE
TOP: This Tiger of schwere Panzer-Abteilung 503 still has smoke candles in the dischargers on the turret sides, revealing that the photo was taken before June 1943. (SB)

OPPOSITE
BOTTOM: The unit emblem (a tiger's head) of schwere Panzer-Abteilung 503 is stenciled onto the right front of the superstructure front plate. A second stowage bin has been added to the right turret side of this Tiger (produced in December 1942 with the Pz.Kpfw.III style stowage bin), which has apparently become immobilized in a gully. (SB)

Tiger (tactical number 241) of the 2.Kompanie/Panzer-Abteilung 503 advancing during Operation Zitadelle. It is one of the survivors from the earlier fighting in January through March 1943. (BA)

A demonstration of a Tiger (tactical number 313 from the 3.Kompanie/schwere Panzer-Abteilung 503) crossing an anti-tank ditch prior to the Kursk offensive. (BA)

Tigers of the 3.Kompanie/schwere Panzer-Abteilung 503 moving up to the front. The "13" stenciled on the underside of the commander's cupola hatch lid was for the 13th step in the procedure for sealing the tank for submerged fording. (CHY)

THIS PAGE AND OPPOSITE: Three views of Tiger (tactical number 334 produced in March/April 1943) of the 3.Kompanie/Panzer-Abteilung 503 which became stuck in the soft ground bordering a stream. (BA)

This Tiger (tactical number 332 of the 3.Kompanie/Panzer-Abteilung 503) became stuck while attempting to cross a marshy stream and is being pulled out by a second Tiger. (BA)

Originally, the schwere Kompanie of SS-Panzer-Regiment 1 of the SS-Division "LSSAH" was number 4, as shown by the tactical number 411 for the platoon leader of the 1.Zug. A large improvised stowage bin was mounted on the turret rear of this Tiger produced in December 1942. (NA)

A closeup of the commander exposed in the open commander's cupola hatch using binoculars to carefully scout the far horizon. The peg on cupola ring behind his back was intended as a mount for a rain shield. A closeup of the hull deck shows the stowage for two gun-cleaning rods and the tow cable. (NA)

The crew of this Tiger from the schwere Kompanie/SS-Panzer-Regiment 1 are assembling five sections of the gun-cleaning rod. (NA)

The rod across the hull front for holding spare track links was welded on in the field by the Werkstatt-Zug and was not an official modification. This Tiger produced in December 1942 still has the K.F.F.2 (double periscopic vision device) for the driver. (NA)

Loading 8.8 cm Sprenggranaten (high explosive shells) with adjustable nose fuses which could be set on delay for penetrating gun shields and firing to ricochet off the ground to produce air bursts. (HLD)

OPPOSITE: General Guderian – Generalinspekteur der Panzertruppen – being shown the Panzerbefehlswagen (command tank) Tiger (tactical number 405) of the schwere Kompanie/SS-Panzer-Regiment 1. The "key" emblem for the unit is stenciled onto the top left corner of the superstructure front plate. To make room for the additional radio set in the turret, the coaxial machinegun wasn't installed and the hole in the gun mantle was plugged. (PA)

Tiger (tactical number 822) of the schwere Kompanie/SS-Panzer-Regiment 2 of SS-Division "Das Reich" being towed by a train of three 18 ton Zugkraftwagen (Sd.Kfz.9). Produced in January 1943, this Tiger has standard sheet metal guards for the exhaust mufflers and S-Mine dischargers mounted on the deck. (BA)

Another Tiger (tactical number 823, renumbered from 832) of the schwere Kompanie/SS-Panzer-Regiment 2 firing to zero the gun sights. Produced in January 1943, it has the five S-Mine dischargers and an improvised stowage bin on the turret rear in place of the Pz.Kpfw.III style. (BA)

A column of Tigers from the schwere Kompanie/SS-Panzer-Regiment 2 (still designated the 8.Kompanie) with the crossed double bar recognition symbols for "Das Reich" during Operation Zitadelle. The new recognition symbols were stenciled on directly before the start of the offensive on 5 July 1943. (BA)

A Tiger from the 9.schwere Kompanie/SS-Panzer-Regiment 3 with the triple bar recognition symbol stenciled on the right of the superstructure front plate. The tactical symbols were to be stenciled in white on the Panzers painted Gruen (green) and in black on the Panzers painted Hellgelb (light yellow). (NA)

Loading 8.8 cm Panzergranaten 39 (APCBC-HE – armor-piercing capped with ballistic cap and a high explosive filler) on a Tiger during Operation Zitadelle in July 1943. (BA)

A large caliber shell fragmented, creating only superficial damage on the face-hardened 100 mm thick superstructure front plate of this Tiger during Operation Zitadelle in July 1943. (PA)

Tigers from the schwere Kompanie/SS-Panzer-Regiment 2 with aerial recognition flags draped across the stowage bins on the back of the turret. The divisional recognition symbol was stenciled on the right rear track guard. (BA)

Tiger (tactical number S11) from the schwere Kompanie/SS-Panzer-Regiment 2 firing high-explosive shells at a fairly close target. (NA)

The commander of Tiger (tactical number S13) from the schwere Kompanie/SS-Panzer-Regiment 2 coordinating actions with infantry. (NA)

This Tiger (tactical number S34 – one of the replacements received by the unit) from the schwere Kompanie/SS-Panzer-Regiment 2 was probably destroyed by the unit by setting off internal charges to prevent an intact Tiger from falling into enemy hands. (PA)

Pz.Kpfw.Tiger Ausf.E Fgst.Nr.250159 completed in March 1943. The unit has sprayed camouflage paint onto the gun barrel which would burn and turn black after rapid firing. (MJ)

ABOVE AND BELOW: A few of the surviving battle-scarred but still operational Tigers (tactical numbers S13 and S33) from the schwere Kompanie/SS-Panzer-Regiment 2 moving on a road march in the late Fall of 1943. (BA)

When sent to the Eastern Front in early May 1943, schwere Panzer-Abteilung 505 was outfitted with 20 Tigers and 24 Pz.Kpfw.III Ausf.M. (BA)

OPPOSITE: A Tiger from schwere Panzer-Abteilung 505 in the Spring of 1943. Barbed wire has been fastened along the superstructure sides to discourage infantry tank-hunter teams. (HLD)

Two Tigers (tactical numbers 321 and 322) of the 3.Kompanie/schwere Panzer-Abteilung 505 which joined the rest of the battalion at the front during Operation Zitadelle. (PA)

The platoon leader's Tiger in the 2.Zug of the 1.Kompanie/schwere Panzer-Abteilung 505 was with Heeres Gruppe Mitte north of the Kursk salient during Operation Zitadelle. (PA)

OPPOSITE TOP: A Tiger captured by the Russians during Operation Zitadelle in July 1943. The gun couldn't be fired with the shattered muzzle brake which provided a high percentage of the recoil-dampening effect. (HLD) BELOW: The external armor cap for the ball machine gun mount on the superstructure front is missing. The tactical number "300" for the company commander of the third company in the battalion has been repainted over a previous number. (KHM)

With no visible tactical numbers, the identity of this Tiger's unit is not known. The 503rd, 505th, and Grossdeutschland were all in the habit of carrying unditching logs on the superstructure side. (BA)

THIS PAGE AND OPPOSITE: These Tigers were completed by Henschel in late April/early May 1943. Some have the loader's periscope but with two track links on the left turret side; others have three track links and the loader's periscope. They are being used to practice platoon and company tactics prior to the unit's transfer to the Eastern Front. (BA)

ABOVE AND OPPOSITE: Tiger crews practice firing on a machine gun range. The cover is securely fastened on the turret ventilation fan, which would not be advisable when firing the 8.8 cm Kw.K.36 L/56. (BA)

PAGES 76-79: Six views of Pz.Kpfw.Tiger Ausf.E (Fgst.Nr.250234, the fourth completed by Henschel in May 1943) with a loader's periscope on the turret roof, spare track link holders on the turret sides, a machined drive sprocket wheel hub, and 12 bolts retaining the steel rim on each roadwheel. It has a turret with welded loader's hatch and reinforced gun mantle by the gun sight apertures. These are propaganda photos taken at the Henschel assembly plant. Just like all other Panzers, Tigers were outfitted at a Heeres Zeugamt (ordnance depot) before being released to a unit. (HLD)

The Pz.Kpfw.Tiger Ausf.E in the foreground with three spare track link holders on the right turret side has already received a base coat of Dunkelgelb – RAL 7028 (dark yellow) paint, while Fgst.Nr.250238 (the 8th completed by Henschel in May 1943 just leaving the assembly hall for acceptance tests) is still coated with the red oxide primer – RAL 8012. (HLD)

PAGES 81-82: This Tiger (tactical number 312) belonging to the 3.Kompanie/schwere Panzer-Abteilung 502 is pulling into a rear area to reload ammunition. It is Fgst.Nr.250235 (the 5th completed by Henschel in May 1943) and has a machined hub for the drive sprocket wheel on the left side but the older style hub drilled for each mounting bolt on the right side. (BA)

PAGES 83-85: Loading 8.8 cm Sprenggranaten (high-explosive shells) on Tiger "312" of the 3.Kompanie/schwere Panzer-Abteilung 502. Even though most cartridge casings were made out of steel, they were electroplated with brass to prevent oxidation. There are still smoke candles in the dischargers on the turret sides, even though this practice had been dropped because hits from small arms set off the smoke candles, which could incapacitate the crew. (BA)

THIS PAGE AND OPPOSITE: Four views of Tiger (tactical number 311) pulling into a rear rest area. The "Mammut" unit symbol is stenciled onto the right front of the superstructure front plate. This and 31 other Tigers with the Stab, 2. and 3.Kompanie of schwere Panzer-Abteilung 502 joined the 1.Kompanie at the front in July 1943. (BA)

OPPOSITE TOP: The fourth Tiger from the 1.Zug of the 3.Kompanie/schwere Panzer-Abteilung 502. These are unique in having an oversize national identity cross stenciled toward the rear on both sides and the tactical numbers stenciled on the hull sides as well as on the turret. (PA)

OPPOSITE BELOW: The adapter with a single hole for the auxiliary engine crank starter on this Tiger of schwere Panzer-Abteilung 502 reveals that it was still one of the first 250 with a Maybach HL 230 P45 engine. The unit removed the Feifel air cleaners. (BA)

ABOVE: The Stab, 10. and 11.Kompanie/III.Abteilung/Panzer-Regiment Grossdeutschland with 31 Tigers joined the rest of the regiment at the front in August 1943. The 10.Kompanie used a tactical numbering system beginning with the letter "B" and the 11.Kompanie used the letter "C". (BA) BELOW: Tiger "314" of schwere Panzer-Abteilung 502 moving out in front of an artillery barrage. (PA)

Tiger (tactical number 231) belonging to the platoon leader of the 3.Zug of the 2.Kompanie/schwere Panzer-Abteilung 502 on display at a demonstration of heavy anti-tank weapons in Heeres Gruppe Nord (army group north). (BA)

Schwere Panzer-Abteilung 506 was the fourth independent heavy tank battalion sent to the Eastern Front. The first unit to be sent to the front with 45 Tigers, it joined Heeres Gruppe Sued in September 1943. (BA)

OPPOSITE: The crew from another Tiger in schwere Panzer-Abteilung 502 taking the time to shave in the morning at their rear rest area. The tree limbs were quite ineffective as camouflage when the gun barrel still prominently protrudes. (BA)

An older Tiger (survivor of Operation "Zitadelle" in the 13.Kompanie, which was renamed as the 9.Kompanie). The 9.Kompanie/Panzer-Regiment Grossdeutschland used a tactical numbering system beginning with the letter "A". (BA)

OPPOSITE: Two Tigers in schwere Panzer-Abteilung 506. The lead Tiger is the third Befehlswagen (command tank) of the battalion identified by the tactical letter "C" and the Sternantenne. The second Tiger with "11" on the turret is the platoon leader's Tiger in the 3.Zug of the 2.Kompanie. (BA)

ABOVE: Each company in schwere Panzer-Abteilung 506 painted the tactical numbers in a different color and numbered their tanks from 1 through 14. (PA)

OPPOSITE BELOW AND ABOVE: Starting in July 1943, the Pz.Kpfw.Tiger Ausf.E was completed with a redesigned turret with seven periscopes in a cast armor cupola. Henschel started applying the Zimmerit anti-magnetic coating in August 1943. (BA)

Schwere Panzer-Abteilung 509 was sent to the Eastern Front with 45 Tigers arriving early in November 1943. The tactical numbers were stenciled on the sides of the turret of their Tigers in stylized font. (PA)

These two Tigers were completed in October 1943 and issued to the schwere Panzer-Abteilung 501. They still have fittings for the Feifel air cleaners on the rear deck and pistol port plugs in the left turret side. (WS)

Schwere Panzer-Abteilung 501 was sent with 45 Tigers to the Eastern Front in December 1943. Two Tigers of the 2.Kompanie are advancing across the bleak landscape in the Winter of 1943/1944. (BA)

The turret of a Panzerbefehlswagen Tiger Ausf.E (Sd.Kfz.267) with the 2 meter long rod antenna for the Fu 5 radio set mounted on the turret roof and the Sternantenne for the long range Fu 8 radio set mounted on the top of the hull on the right side. (HLD)

Tigers of the schwere Panzer-Abteilung 507 in an assembly area preparing for an attack on Tarnopol in April 1944. The Befehlswagen (command tank) in the battalion headquarters company were identified by the letters A, B, and C. Other Tigers in the battalion were identified by a larger first digit in their tactical number stenciled on the turret side. (BA)

The Befehlswagen (command tank) for the battalion commander of schwere Panzer-Abteilung 507 was identified by the letter "A". (BA)

A column of Tigers from schwere Panzer-Abteilung 506 advancing to attack Tarnopol in April 1944. The second Tiger forward from the rear of the column is a Befehlswagen with the tube for stowing the spare antenna rods mounted across the hull rear. (BA)

Starting with Fgst.Nr.250634, Tigers (such as this one completed by Henschel in December 1943) had a rear travel lock for the 8.8 cm Kw.K.36 gun. (BA)

A Bilstein 3 ton crane mounted on a truck is being used to lift a Maybach HL 230 P45 engine. (BA)

THIS PAGE AND OPPOSITE: This Pz.Kpfw.Tiger Ausf.E no longer has the mountings for the Feifel air cleaners on the hull rear and had a 20-ton jack which Henschel started mounting on Tigers completed in the latter half of January 1944. With the exception of one section of the track guard and the area around the national identity cross, it appears that this Tiger has been coated with whitewash for winter camouflage. (BA)

Tigers crossing a makeshift bridge over a frozen stream. If the driver maintained an even slow speed without shifting gears, a combat-loaded 57 metric ton Tiger could cross bridges designed to carry 18 metric tons. (PA)

OPPOSITE:
TOP: A survivor of almost one year at the front, this Tiger in the 1.Kompanie/schwere Panzer-Abteilung 503 is moving out from the assembly area with Panzer-Grenadiere riding on the rear deck in December 1943. (BA)

BOTTOM: A Tiger of the 2.Kompanie/schwere Panzer-Abteilung 503 being towed into position under the portal crane by another Tiger. (SB)

PAGES 106-109: Another survivor of about a year in combat on the Eastern Front, a Tiger covered with whitewash as winter camouflage in the schwere Panzer-Abteilung 502 is towing another Tiger across a shallow ford. (BA)

THIS PAGE AND OPPOSITE: : Since the 502nd was overstrength by 26 Tigers at the end of February 1944, additional Tigers were also assigned to the headquarters platoon of each company of schwere Panzer-Abteilung 502. (2-BA, 2-PA)

ABOVE AND OPPOSITE: One of the 12 replacement Tigers that arrived in April 1944 for schwere Panzer-Abteilung 505 was assigned to the company commander of the 3.Kompanie. It is Fgst.Nr.250829, completed by Henschel in early February as one of the first with steel-tired rubber-cushioned roadwheels. (BA)

Schwere Panzer-Abteilung 502 had received 52 replacement Tigers in January and February 1944, building them up to a peak strength of 71 Tigers on 29 February 1944. Therefore, there were more than four Tigers assigned per platoon, as shown by this 7th Tiger in the 1.Zug of the 2.Kompanie. (BA)

Another of the replacement Tigers for schwere Panzer-Abteilung 505 with the unit symbol (a charging knight) stenciled on the turret side and the tactical number stenciled onto the armor recoil guard. (BA)

This Tiger in Hungarian service was completed late in February 1944 after introduction of the turret ring guard and after the external travel lock on the hull rear was dropped. (GF)

A Tiger in an assembly area being screened from view from the side by a row of branches struck in the ground. (GF)

Two Tigers from schwere Panzer-Abteilung 503 advancing across a wide river valley in the Spring of 1944. The closer Tiger (tactical number 312) has survived about two years in combat. (GF)

Tiger "A22" from the III.Abteilung/Panzer-Regiment Grossdeutschland loaded with the narrow transport tracks on a special SSYMS railcar. (BA)

Another replacement Tiger "B12" in the 10.Kompanie/III./Panzer-Regiment Grossdeutchland moving past several captured Russian 76.2 mm field guns. (BA)

Replacement Tiger "A12" in Panzer-Regiment Grossdeutschland changing from transport to cross-country tracks. Produced after April 1944, it has the monocular gun sight and thicker (40 mm) turret roof. (BA)

As a replacement for a Befehlswagen, this normal Pz.Kpfw.Tiger Ausf.E has been assigned to the battalion commander's tank (tactical number S01) of the III.Abteilung/Panzer-Regiment Grossdeutschland. (BA)

In an attempt to improve protection, additional spare track links were carried on the right turret side of this Tiger I in the 3.Kompanie/schwere Panzer-Abteilung 507. (PA)

One of the replacements sent to the 9.schwere Kompanie/SS-Panzer-Regiment 3 in the Summer of 1944. This was the only heavy company in the SS that remained with the Panzer-Regiment; the other two were incorporated into independent schwere SS-Panzer-Abteilungen. (BA)

Schwere Panzer-Abteilung 510 was the last independent army heavy tank battalion sent to the Eastern Front with 45 Tiger I in the Summer of 1944. In this staged attack, the Tiger is escorting Fallschirmjaeger (parachute troops) across an open field in Lithuania. (CHY)

Loading ammunition and refueling another Tiger I of schwere Panzer-Abteilung 510 in August 1944. (BA)

OPPOSITE:
TOP: Welding holders for the spare track links to the front hull locally weakened the frontal armor of this Tiger I in schwere Panzer-Abteilung 510. (PA)

BOTTOM: A replacement Tiger I for schwere Panzer-Abteilung 509 is unique in having a Nahverteidigungswaffe (close defense weapon) mounted in a 25 mm thick turret roof. It also has a field modification where a sheet metal strip has been welded across the front edge of the turret roof to provide a rain guard over the gap behind the gun mantlet. (PA)

A Tiger I completed by Henschel in April/May 1943, claimed to have been knocked out by the Russians on the Eastern Front in 1945. (PA)

TIGER I IN ITALY

PAGES 125-126: This Tiger (tactical number 8) nicknamed "Strolch" was one of the eight in Tigergruppe Meyer sent to Italy in August 1943. (BA)

Schwere Panzer-Abteilung 508 was sent down to Italy with 45 Tigers to attempt to wipe out the Allied bridgehead at Anzio. (BA)

Frequent maintenance halts had to be made during long road marches in order to keep the Tigers operational. Because of their haste and the mountainous terrain, schwere Panzer-Abteilung 508 lost over 50 percent of its Tigers to mechanical breakdowns on the march from the railhead to the front. (BA)

ABOVE AND OPPOSITE: A broken track and damaged roadwheels are the result of hitting a mine. Most of the first 45 Tigers in schwere Panzer-Abteilung 508 were completed in November/December 1943. (BA)

A Tiger belonging to schwere Panzer-Abteilung 508 being towed by two 18 ton Zugkraftwagen (Sd.Kfz.9). (BA)

This is not a Bergetiger!. As reported by 14.Armee on 17 April 1944, schwere Panzer-Abteilung 508 was already busy designing a Minenraeumgeraet (mine clearing device) for a Tiger. The gun was removed and a boom socket with a hand-cranked winch were mounted on the turret as a means of laying explosive charges to detonate mines. (TTM)

A replacement Tiger for the 3.Kompanie/schwere Panzer-Abteilung 508 had taken four hits by 75 mm or 76 mm armor piercing shells from angles of 30 to 60 degrees from its right rear. All failed to penetrate. (MJ)

PAGES 132-135: Replacement Tigers for the 1.Kompanie/schwere Panzer-Abteilung 508. One still has the binocular sight and 25 mm thick turret roof while the other has a monocular sight and 40 mm thick turret roof. (BA)

A Tiger from schwere Panzer-Abteilung 508 abandoned near Cisterna, Italy in May 1944. (NA)

Another Tiger from schwere Panzer-Abteilung 508 abandoned after it ran into a shell or bomb crater. (NA)

A Tiger in the 3.Kompanie/schwere Panzer-Abteilung 504 sent down to Italy in early June 1944 to halt the Allied drive. (MJ)

Having given up the attempt to tow this Tiger from schwere Panzer-Abteilung 508, the crew set it on fire (all rubber missing from the roadwheels) to prevent it being captured intact. (TTM)

ABOVE AND OPPOSITE: After taking a hit though the superstructure side into the fuel tanks and engine compartment, this Tiger was demolished by the crew. The hydraulic fluid was drained out of the recoil cylinder and a round fired to destroy the gun. The ammunition was detonated, blowing out the pannier on the right side as well as bulging and cracking the hull roof. (NA)

Another Tiger abandoned after being prepared for towing by shortening the track to fit around the lead roadwheels on the left side. Exploding ammunition (probably set off by internal demolition charges) has blown out the hull roof along the right side. (NA)

TIGER I ON THE WESTERN FRONT

Nineteen of the 45 Tiger I for schwere SS-Panzer-Abteilung 101 were acquired by the unit in 1943 and still had the dished roadwheels with rubber tires. (BA)

Completed in December 1943, this Tiger (tactical number 321) issued to schwere SS-Panzer-Abteilung 101 has a rear travel lock and has the headlight mounted in the center of the superstructure front plate (a modification initiated in October 1943). (BA)

Tiger I (tactical number 331) is also one of the 19 Tiger I issued to schwere SS-Panzer-Abteilung 101 in 1943. The external travel lock can clearly be seen in silhouette. (BA)

These Tiger I from the 2.Kompanie are several of the 26 Tiger I issued to schwere SS-Panzer-Abteilung 101 in April 1944. They have the rubber-cushioned steel-tired roadwheels and monocular gun sights. (BA)

A Tiger I in the 2.Kompanie/schwere SS-Panzer-Abteilung 101 still has a binocular gun sight. The thickness of the turret roof has been increased to 40 mm as a defense against heavier caliber (150 mm) high-explosive shells. (BA)

The frontal armor of this Tiger I of schwere SS-Panzer-Abteilung 101 successfully withstood the punishment from the opponent's anti-tank guns. (NA)

OPPOSITE:
TOP: Film footage of Michael Wittman in the commander's cupola of a Tiger I camouflaged with tree branches. (NA)

BOTTOM: A Tiger I from schwere SS-Panzer-Abteilung 101 knocked out in Villers Bocage. (BA)

ABOVE AND OPPOSITE: Armor piercing shot fired from a 17-pounder gun successfully penetrated the turret side of this Tiger I. An unsuccessful shot is lodged in the 100 mm thick superstructure frontal armor. (2 TTM, 1 PAC)

Two views of three damaged Tiger I that had been loaded on rail cars for evacuation. They were captured by U.S. troops in Rheims, France on 31 August 1944. (NA)

Two Tiger I from Panzer-Abteilung (Fkl) 301 fitted with the narrower loading tracks for transport on SSYMS heavy duty rail cars. (MJ)

During the Battle of the Bulge, this Tiger from the 4.Kompanie/schwere Panzer-Abteilung 506 was captured near Oberwampach in Luxembourg in January 1945. (NA)

ABOVE AND OPPOSITE: Issued to the last unit outfitted with Tiger I sent to the Western Front, Panzer-Abteilung (Fkl) 301, this Tiger I completed in August 1943 has factory-applied Zimmerit anti-magnetic coating. (MJ)

A Tiger from the 1.Kompanie/Panzer-Abteilung (Fkl) 301 which were outfitted with an additional radio set to steer "B IV c – schwere Ladungstraeger" (heavy high-explosive charge carriers) onto targets. (MJ)

TIGER II ON THE WESTERN FRONT

One of the first five production Tiger II (Fgst.Nr.280001 to 280005) in the 1.schwere Panzer-Kompanie (Fkl) of the Panzer-Lehr Division. (MJ)

OPPOSITE:
TOP: Pz.Kpfw.Tiger Ausf.B (Fgst.Nr.280008) completed by Henschel in February 1944 and issued to the testing facility in Kummersdorf. It has Zimmerit (anti-magnetic coating), the curved exhaust pipes, and the Gg 24/800/300 cross-country tracks. (BA)

BOTTOM: Details of the drain covers and escape hatch on the bottom of the hull are seen in this photo of a Pz.Kpfw.Tiger Ausf.B being paid a visit by General Eisenhower. The trailing roadwheel suspension arms on the right side and leading roadwheel suspension arms on the left side are also revealed. (NA)

ABOVE AND OPPOSITE: Pz.Kpfw.Tiger Ausf.B (Fgst.Nr.280006) completed by Henschel in February 1944 was issued to the Panzerversuchsstelle (tank testing facility) in Kummersdorf. It was outfitted with the monoblock 8.8 cm Kw.K.43 L/71 gun with the heavy muzzle brake and the binocular gun sight. As with all Tigers produced up to mid-August 1944, it left the assembly plant with a base coat of Dunkelgelb RAL 7028 (dark yellow) paint applied over the entire outer surface, including the Zimmerit. (BA)

PAGES 160-163: Two of the 12 Tiger II in the 1.Kompanie/schwere Panzer-Abteilung 503 keeping out of sight of Allied aircraft. Both Tiger II have sectional gun tubes and monocular gun sights. (BA)

PAGES 164-166: A Tiger II from the 1.Kompanie/schwere Panzer-Abteilung 503 knocked out in a village in France. The displaced turret provided a clear view of the segmented turret ring guard bolted to the top of the hull. This Tiger II has a monoblock gun tube, monocular gun sight, and double section track. (TTM)

PAGES 167-168: These three Tiger II were issued to the 1.Kompanie/schwere Panzer-Abteilung 503 and captured by the British in Normandy. Completed by Henschel in April and May 1944, they feature the sectional 8.8 cm Kw.K.42 L/71 guns with the heavier muzzle brake, monocular gun sights, drive sprockets with nine teeth for the double link Gg 26/800/300 cross-country track with its single piece connecting link. (TTM)

PAGES 169-173: Three Tiger II (tactical numbers 313, 323, and 324) in training with the 3.Kompanie/schwere Panzer-Abteilung 503. The segmented turret ring guard can be seen on the closeup of "323". All but a few of the 14 Tiger II issued to the 3.Kompanie still had the so-called "Porsche-Turm" designed by Krupp. (BA)

PAGES 174-177: Four views of the maintenance personnel from schwere Panzer-Abteilung 503 spray painting camouflage patterns onto this Tiger II with a Serienturm designed by Krupp. It was one of the few Tiger II with a Serienturm that didn't have spare track link hangers welded to the turret sides. (BA)

Remounting the track guards after spray painting the camouflage pattern onto this Tiger II (tactical number 300), which belongs to the company commander of the 3.Kompanie/schwere Panzer-Abteilung 503. (BA)

Varying camouflage patterns were applied by the unit, as is evident in photos like this one of a Tiger II (tactical number 332) issued to the 3.Kompanie/schwere Panzer-Abteilung 503. Stripes of Olivgruen RAL 6003 and Rotbraun RAL 8017 had been spray painted over the base coat of Dunkelgelb RAL 7028. (BA)

Strikes from armor-piercing projectiles on the thick sloped frontal armor of this Tiger II barely dented the surface. (APG)

THIS PAGE AND OPPOSITE: Pz.Kpfw.Tiger Ausf.B (Fgst.Nr.280093) completed by Henschel in July 1944 and issued to the 1.Kompanie/schwere SS-Panzer-Abteilung 101. It is now on display in England. It has an 40 mm thick loader's hatch and welded base commander's cupola. (TTM)

Tiger II (Fgst.Nr.280101), completed by Henschel in July 1944, was issued to the 1.Kompanie/schwere SS-Panzer-Abteilung 101. It was found upside down with the end of the barrel blown off. Sent back to Aberdeen Proving Ground for study, it was later returned to Germany where it is now on display at the Panzer Museum in Munster. (APG)

The turret on Pz.Kpfw.Tiger Ausf.B (Fgst.Nr.280105), completed by Henschel in late July 1944, still has the 15 mm thick forged loader's hatch, three Pilze (sockets) for mounting the 2 ton jib boom, and lacks armor guards for the torsion bars counterbalancing the turret rear hatch. (NA)

Two Tiger II in schwere Panzer-Abteilung 506 concealed in a fruit orchard in the Aachen area in October 1944. (BA)

THIS PAGE AND OPPOSITE: A Tiger II from schwere Panzer-Abteilung 506 that was stopped by hits in the track and drive sprocket and then finished off with a penetration through the turret side. The kill was claimed by the 702nd Tank Destroyer Battalion of the 2nd Armored Division on 28 November 1944. (NA)

THIS PAGE AND OPPOSITE: Tiger II (tactical number 2+11) from the 2.Kompanie/schwere Panzer-Abteilung 506 was captured by American troops and restored to running condition by Company B, 129th Ordnance Battalion, by 15 December 1944. (NA)

A replacement Tiger II for Panzer-Abteilung 506 captured by the Americans in the Winter of 1944/45. (NA)

PAGES 189-190: Three of the 45 Tiger II in schwere SS-Panzer-Abteilung 501 advancing at the start of operation "Wacht am Rhein", commonly known as the Battle of the Bulge. Henschel ceased applying coats of Zimmerit on new production Tigers after receiving an order to stop dated 8 September 1944. (NA)

PAGES 191-193: Tiger II (Fgst.Nr. 280243, tactical number 332), completed by Henschel in September 1944, was first issued to schwere Panzer-Abteilung 509, which subsequently turned it over to schwere SS-Panzer-Abteilung 501. Captured by the Americans, it was shipped to Aberdeen Proving Ground for study. It is currently on display at the Patton Museum in Fort Knox, Kentucky. (NA)

Tiger II (tactical number 204) belonging to the company commander of the 2.Kompanie/schwere SS-Panzer-Abteilung 501 was abandoned near La Gleize, Belgium before 4 January 1945. The camouflage pattern of dots and swirls was applied at Henschel using well-thinned Dunkelgelb RAL 7028 (dark yellow) paint covering less than two-thirds of the armor surface, which was only covered by a base coat of dull red oxide primer. Captured in running condition, it subsequently broke down when the Americans tried to drive it to a railhead. (NA)

One of the 21 Tiger II taken into action by schwere Panzer-Abteilung 507 at the end of March 1945. This Tiger II received a base coat of Olivgruen RAL 6003 (olive green) paint at Henschel, with Dunkelgelb RAL 7028 sparsely applied in dots and swirls to create the camouflage pattern. (NA)

A Tiger II abandoned in early April 1945 near Mahmecke, Germany which Combat Command R of the 7th U.S. Armored Division claimed to have knocked out in the division's attack in the Ruhr pocket. (NA)

THIS PAGE AND OPPOSITE: Four additional Tiger IIs abandoned by the Germans as they pulled back after their offensive failed during the Battle of the Bulge. There are no signs of penetrations through the armor. One has been destroyed by setting off internal demolition charges which blew out the right superstructure and hull side. The 8.8 cm Kw.K.43 L/71 guns were also disabled by firing a round after draining the hydraulic fluid from the recoil cylinders. (NA)

A Tiger II from schwere Panzer-Abteilung 507 completed at Henschel in March 1945 has 18 teeth in each drive sprocket for the new single section Kgs 73/800/152 track. This Tiger II also received a base coat of Olivgruen RAL 6003 (olive green) paint at Henschel, with Dunkelgelb RAL 7028 sparsely applied in dots and swirls to create the camouflage pattern. (NA)

TIGER II ON THE EASTERN FRONT

A very rare shot of the Behelfskran 2t (jib boom) mounted in the three Pilze (sockets) on the turret roof of this Tiger II from schwere Panzer-Abteilung 501 being used to remove the rear deck over the engine compartment. The loader's hatch is still the 14 mm thick forging. (BA)

A Tiger II in the 3.Kompanie/schwere Panzer-Abteilung 501 accompanied by two Sturmgeschuetz, engaged in mopping up a bridgehead in the Weichsel bend on 2 October 1944. (BA)

A Tiger II (tactical number 324) in the 3.Kompanie/schwere Panzer-Abteilung 501, which was the first unit sent to the Eastern Front with 45 Tiger II. (MJ)

Two views of one of the three Panzerbefehlswagen Tiger Ausf.B issued to schwere Panzer-Abteilung 501 after it was captured by the Russians. It was outfitted as an Sd.Kfz.267 with a long range Fu 8 radio set. The Sternantenne was to be mounted on a flexible base on top of a white porcelain insulator protected by an armor cylinder centered at the end of the rear deck (not as shown). (PA)

After losing all of their Tigers in France by the end of August 1944, the schwere Panzer-Abteilung 503 was outfitted with 45 Tiger II in the Fall of 1944. Here the 3.Kompanie is being paraded to create a propaganda film. (NA)

PAGES 204-206: Tiger II (tactical number 233) in the 2.Kompanie/schwere Panzer-Abteilung 503 in Budapest in October 1944. (BA)

A closeup of the turret front of a Tiger II completed by Henschel in late August/early September 1944 and issued to schwere Panzer-Abteilung 503. Apparently, the commander didn't like to fight buttoned up, as the hatch lid for the commander's cupola has been removed from the pivoting arm. (BA)

A crew member is tying an impregnated canvas bag over the muzzle brake of the 8.8 cm Kw.K.43 L/71 gun of this Tiger II in schwere Panzer-Abteilung 503 in Budapest, Hungary in October 1944. (BA)

This Tiger II, belonging to the company commander of the 2.Kompanie/schwere Panzer-Abteilung 503, crawls over a rubble barricade in Budapest in October 1944. (BA)